A Guide to Growing Pineapples under Glass

by

David Thomson

Copyright © 2013 Read Books Ltd.
This book is copyright and may not be
reproduced or copied in any way without
the express permission of the publisher in writing

British Library Cataloguing-in-Publication Data
A catalogue record for this book is available from the
British Library

Fruit Growing

In botany, a fruit is a part of a flowering plant that derives from specific tissues of the flower, one or more ovaries, and in some cases accessory tissues. In common language use though, 'fruit' normally means the fleshy seed-associated structures of a plant that are sweet or sour, and edible in the raw state, such as apples, oranges, grapes, strawberries, bananas, and lemons. Many fruit bearing plants have grown alongside the movements of humans and animals in a symbiotic relationship, as a means for seed dispersal and nutrition respectively. In fact, humans and many animals have become dependent on fruits as a source of food. Fruits account for a substantial fraction of the world's agricultural output, and some (such as the apple and the pomegranate) have acquired extensive cultural and symbolic meanings. Today, most fruit is produced using traditional farming practices, in large orchards or plantations, utilising pesticides and often the employment of hundreds of workers. However, the yield of fruit from organic farming is growing – and, importantly, many individuals are starting to grow their own fruits and vegetables. This historic and incredibly important foodstuff is gradually making a come-back into the individual garden.

The scientific study and cultivation of fruits is called 'pomology', and this branch of methodology divides fruits into groups based on plant morphology and anatomy. Some of these useful subdivisions broadly incorporate 'Pome Fruits', including apples and pears, and 'Stone Fruits' so called because of their characteristic middle, including peaches, almonds, apricots, plums and cherries. Many hundreds of fruits, including fleshy fruits like apple, peach, pear, kiwifruit, watermelon and mango are commercially valuable as human food, eaten both fresh and as jams, marmalade and other preserves, as well as in other recipes. Because fruits have been such a major part of the human diet, different cultures have developed many varying uses for fruits, which often do not revolve around eating. Many dry fruits are used as decorations or in dried flower arrangements, such as lotus, wheat, annual honesty and milkweed, whilst ornamental trees and shrubs are often cultivated for their colourful fruits (including holly, pyracantha, viburnum, skimmia, beautyberry and cotoneaster).

These widespread uses, practical as well as edible, make fruits a perfect thing to grow at home; and dependent on location and climate – they can be very low-maintenance crops. One of the most common fruits found in the British countryside (and towns for that matter) is the blackberry bush, which thrives in most soils – apart from those which are poorly drained or mostly made of dry or sandy soil. Apple trees are, of course, are another classic and whilst they may take several years to grow into a well-established tree, they will grow nicely in most sunny and well composted areas. Growing one's own fresh, juicy tomatoes is one of the great pleasures of summer gardening, and even if the gardener doesn't have room for rows of plants, pots or hanging baskets are a fantastic solution. The types, methods and approaches to growing fruit are myriad, and far too numerous to be discussed in any detail here, but there are always easy ways to get started for the complete novice. We hope that the reader is inspired by this book on fruit and fruit growing – and is encouraged to start, or continue their own cultivations. Good Luck!

THE PINE-APPLE.

THIS noble fruit has derived the name of pine-apple from its striking resemblance in shape to the cones of some of the pine-trees. It is probably the most rich and luscious of fruits. "Three hundred years ago it was described by Jean de Levy, a Huguenot priest, as being of such excellence that the gods might luxuriate upon it, and that it should only be gathered by the hands of a Venus."

Some say that it is a native of Brazil, and found its way from that country to the East. It is, however, not very clearly determined to what part of the world we are indebted for the pine-apple; and there is little doubt that it is also a native of the West Indies, for many of its varieties are found growing wild on the continent and islands of the West. It was first brought into Europe by a Dutch merchant, and introduced into this country from Holland in 1690; and first cultivated for the dessert by Mr Bentinck, ancestor to the present ducal family of Portland.

The superior cultivation of the pine-apple has always been regarded as one of the greatest triumphs of horticulturists. Improved practice is perhaps as much apparent in pine-culture as in any branch of horticulture. Superior results are now attained in eighteen months to what it required twice that time to produce in the recollection of the writer. To Mr James Barnes, late gardener at Bicton Park, Devonshire, we are indebted for exposing and discontinuing the erroneous practice of annually disrooting pine plants, and subjecting them to too high a soil temperature. This was the first step in contracting the period considered necessary to bring the pine-apple to maturity. And of more recent date is the very general cultivation of the pine-apple in much smaller pots than were used some thirty-five years ago: and where the pot system is practised, the use of smaller pots makes them more easily managed, and at less expense.

PINERIES.

That which naturally claims attention first in treating on the cultivation of the pine-apple is, the description of houses or pineries which afford the greatest convenience and facilities for first-rate cultivation, their situation, and the exposure which they should occupy.

The situation should be one well sheltered from cutting winds, and having a full south aspect. There is nothing that necessitates hard firing to keep up a given temperature more than exposure to high winds; and the atmosphere will be the more conducive to healthy growth the less firing is required to maintain the heat. Therefore, shelter from north, east, and west should be taken into consideration in the erection of pineries, especially if the situation is naturally exposed to high winds. It must, however, be borne in mind, that whatever the sheltering objects, they must not be allowed to interfere with full exposure to sunshine at all seasons of the year.

During by far the greater portion of the year, pines cannot possibly have more light and sun than are necessary to produce a stocky fruitful growth in the dull atmosphere which so much prevails in this country. Pineries should therefore be constructed so as to admit and diffuse as much light and sunshine as can be had. In the few months when at times the sun may be more scorching than is desirable, a slight shading can easily be applied. When the sash-and-rafter principle is adopted, I would advise that the sashes should not be less than 6 feet wide, and divided into five openings or panes of glass.

For summer growth I would give the preference to

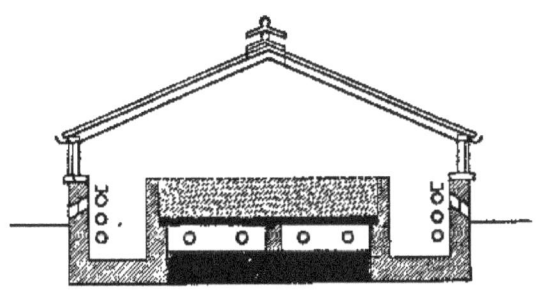

FIG. 1.

span-roofed houses, running north and south (fig. 1). In the morning and afternoon they receive the full sun; and for a period in the middle of the day, when the sun is in meridian, the pines are, in such houses, partially shaded from the scorching rays of the sun, while at the same time they are exposed to a great diffusion of light. Such houses are decidedly the best for summer growth; but, for six months of the year, they do not, from their position, embrace so much direct sunshine as a lean-to house facing due south. Moreover, from the greater amount of glass as a radiating surface in span-roofed houses, they require more fire-heat to keep up the temperature. In these respects the lean-to gives advantages over the span-roofed pinery, in whatever position the latter is placed. For starting pines in December and the two following months, as well as for swelling off fruit during winter and early spring, I recommend lean-to houses, as represented by fig. 2.

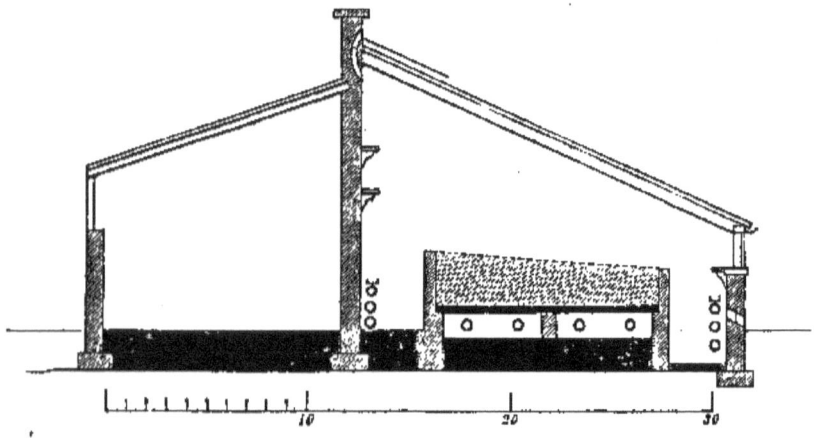

FIG. 2.

The dimensions of the two pineries represented by the woodcuts, are 40 feet by 18 feet, which give a house of handsome proportions. But as the extent of the pineries must be guided entirely by the supply required, I will not enter further into this question. Suffice it to say, that it is more desirable to have several structures of moderate size than a less number of larger ones. A constant succession of ripe fruit is much more easily kept up by having a number of compartments.

For suckers, a common lean-to pit, as represented by fig. 3, is very well adapted, as the young plants can be kept near the glass, and well exposed to light. Where expense is not an object, and for the sake of convenience, this pit may be wider, and have a path along the back, in which case another row of pipes will be necessary.

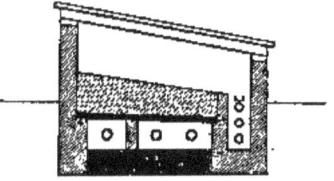

Fig. 3.

But as the woodcuts given will explain more correctly than words the description of pineries recommended, I will not extend my remarks under this heading. It will be observed that the accommodation which I prefer and recommend is partly span-roofed and partly lean-to.

In the formation of the pine ground, the lean-to or early houses should be on the north of the space selected, so that the back affords the shelter from the north which is so desirable; the span-roofed structures to stand north and south, or at right angles with the early lean-to houses, and at a sufficient distance from them not to obstruct sunshine. The early house is thus nearest the boiler in the back shed, and forms the very best shelter to the span-roofed or succession pits, which should not be very high. I am aware, indeed, from experience, that such houses and arrangements are not absolutely necessary for the production of first-rate pines; but they afford great advantages and convenience, and I recommend them as admirably adapted for the culture of this noble fruit.

The pine-apple being a fruit which requires a high temperature, particularly in some of its stages of growth, there should be a good command of heat both for top and bottom. It is not only a false economy to stint the amount of pipes employed, but a larger heating surface moderately heated is much more conducive to the health of plants than a smaller surface kept at scorching heat. I therefore recommend, as shown in the sections given, a liberal amount of pipes and plenty of boiler-power. Besides this I feel fully persuaded, from my experience, that coverings applied to the glass, particularly in the case of fruit swelling off during the colder months of the year, are an immense advantage. A high and steady temperature can be much more easily and economically maintained, and without a parched atmosphere, which in the case of hard forcing in winter requires so much and such constant counteracting.

I have a decided objection to flat-roofed pineries. They are dark, and very productive of drip in winter—conditions the most undesirable in the culture of most plants, and especially so in that of the pineapple. Ventilation should be amply provided for at the apex of the roof; and, particularly in fruiting-houses, there should also be ventilators at intervals along the front, so placed as to cause the air to pass inward in contact with the hot-water pipes. Front ventilation is not to be recommended as a rule; but it is well to provide for it in the erection of pineries, so that in very hot calm days it can be applied, especially in the case of fruit that are colouring.

All pineries and pits should be provided with a steadily-acting steaming apparatus, which can be used or not according as circumstances demand.

A great many methods of supplying moisture to the atmosphere of hothouses have been adopted—such as zinc troughs placed on the pipes, troughs cast on the pipes themselves, a flow of water running in an open gutter, rising out of the flow-pipe at one end of the house and dropping into the return at the other. I have tried all these ways, and more besides, and consider them all inferior to that represented by fig. 4. This is a flat-bottomed open gutter or trough, 6 inches wide, and 2½ inches deep, running the whole length of the house. In the centre and along the whole length of the trough is fixed a rain-water or lead pipe, 2½ inches in diameter. This, as will be seen, is connected with the flow-pipe as it leaves the boiler, and with the return-pipe at the other end of the house. At the middle of the house a tap is fitted into the 2½-inch pipe; a flow of water from the tap can be so adjusted as to let water sufficient trickle into the

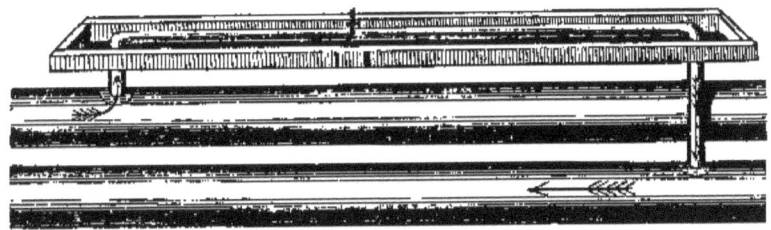

Fig. 4.

trough to keep it full and the small pipe nearly immersed in water. The supply to the boiler being by ball-cock, the small quantity of water that escapes from the tap is constantly supplied. This apparatus requires next to no attention, and heats regularly the whole length of the house. In open gutters without this small pipe, we have always found too much steam at one end of the house and next to none at the other, especially in long houses. The arrangement we recommend is quite equal in heating power to a row of 4-inch pipe. When atmospheric moisture is not required, the water can be dried up out of the trough by simply turning the tap. This system of supplying moisture is applicable in the case of forcing the other fruits treated of in this volume. The pipes should also be so arranged that, by means of stop-cocks, the bottom-heat can be shut off, and applied and regulated according to the amount recommended for the different stages of the growth of the pine.

In all pine-stoves where there is not a supply of soft water from lake or stream, there should be a tank into which to conduct the rain-water from the roof, and passing through the tank a coil of hot-water pipe to warm it. This, in cases where pines are grown extensively, saves a vast amount of trouble in warming water, or in drawing it from the heating apparatus, which latter, for several reasons, is not desirable.

The arrangement of the plants in the various kinds of pineries is a matter worth referring to. In lean-to houses the tallest plants should always be in the back row, and in span-roofed houses they should be placed in the centre row, so that in each case the plants form a sloping bank of foliage all fully exposed to the sun. Where the plants are of very equal growth, the centre of the bed in span-roofed houses should be a little higher.

As I intend to refer to the management of the leaf-and-tan bed in the cultural directions to be given, I will not here enter on that question. I may just state that, apart from the increased labour and liability to violent heating, I have a warm side for the tan-and-leaf bed for pine-growing. I consider the heat derived from this old-fashioned source second to none other for the production of fine pines. Yet I would never prefer it to hot water, because it entails more labour and much more watchfulness, which, in these high-pressure days, is a powerful argument in favour of deriving all the heat from hot water, by which means it can be easily applied and regulated to a degree. Nevertheless, I intend to speak of the management that I adopt in the case of pines grown on a bed of leaves and tan for the supply of bottom-heat. To derive top-heat from fermenting material is a thing which, I believe, is now rarely thought of, and is, to say the least of it, an expensive and cumbrous system.

VARIETIES OF PINES.

In making a selection of varieties, it is not necessary to have many in order to keep up a constant supply of first-rate pines. I believe I am correct in saying that nearly all pine-growers have discontinued the practice of growing so many varieties as were commonly grown many years ago, and will not, therefore, give an extended list, but will enumerate and shortly describe those which are considered the best, and indispensable in pine-growing establishments of ordinary dimensions.

THE QUEEN.—This old and well-known variety still holds its position as one of the best for ripening from May till the end of October. It is a free grower, dwarf and compact in habit, a very certain fruiter, comes quickly to maturity, is very handsome in shape, and of a rich golden colour. Its flavour, as a summer and autumn pine, is not excelled by any other, and it keeps in good condition for three weeks after being ripe. It propagates itself freely by suckers. From May till the end of October there is no pine to surpass it for general excellence; but it will not swell freely in winter, and, as a winter pine, is generally wanting in juiciness and flavour. The Ripley and Moscow Queens are distinct varieties of this, and both good.

SMOOTH-LEAVED CAYENNE.—Taken as a whole, this is the finest pine I know for supplying ripe fruit from October till May, and is the most generally useful variety in cultivation. It swells more freely, and is more juicy in winter, than any other pine that I have grown, and its flavour is excellent. The habit of the plant is somewhat taller than the Queen, and more spreading, with very broad, brittle, dark-green leaves. It is a large and handsome fruit, and, when well swelled, weighs a pound for every pip in depth. Colour a rich yellow, shape slightly conical; when swelled to its best it is rather barrel-shaped. This splendid pine has taken a high position in most collections. For some time spurious smooth-leaved varieties were thrown on the market for this one, and in consequence it fell into considerable disrepute; but it has now fairly established its deservedly high position among pines. It should be in all collections.

BLACK JAMAICA.—Tall and erect in growth, a certain fruiter, medium size, with large flat pips, rather dull in colour, very high flavoured, probably the highest flavoured winter pine in cultivation; but some object to its hardness of flesh, and prefer the Smooth Cayenne on account of the melting juiciness of the latter. Still there can be no doubt of the excellence of the flavour of this variety, and a few of it should be cultivated wherever winter pines are esteemed.

WHITE PROVIDENCE.—A strong and tall-growing variety. Leaves very broad, and covered with down. It yields the largest fruit of any variety in cultivation. Globular in form, with very large flat pips. Flavour quite second-rate. It is an easily-grown and free-fruiting pine; but unless where there is plenty of room it is not to be recommended, and a few plants are sufficient in the largest collection.

CHARLOTTE ROTHSCHILD.—Resembles the Smooth Cayenne in size and habit of plant, but its leaves are studded with strong spines; fruit large, flavour good; is a splendid winter pine—in this respect almost equal to the Cayenne; is a certain fruiter, and grows to a large size. I have ripened it in 11-inch pots, weighing 11 lb. It should be in every collection.

PRINCE ALBERT.—A tall but very compact grower, can be grown in the same space as a Queen. Fruit large, conical, very showy; crown small. Swells well in winter. Flesh soft, very juicy and well flavoured. Free fruiter. It has the fault of not keeping many days after it is ripe, and often large fruits of it begin to decay at their base before they are coloured to the top. A few only should be grown.

LAMBTON CASTLE SEEDLING.—This splendid variety was put into commerce in 1878, and it fully maintains its good character. Remarkable for its free-fruiting habit and large fruit. We believe it is capable of being grown to 12 lb. weight. Fig. 5 is an engraving from a photograph of a fruit ripened in midwinter at Lambton Castle on a plant 19 months old. The fruit measured 12 inches high and 20 inches in circumference, and weighed over 10 lb.; and including the crown, the height from the surface of the pot did not exceed 30 inches. Colour of fruit high orange. Foliage robust, and thinly furnished with unusually strong spines. Keeps well after being ripe, and is exceedingly juicy and well flavoured.

There are a great many more varieties which I might describe, such as different varieties of the Queen, Black

Prince, Enville, Prickly Cayenne, Globe, Antigua, and
Blood-red, &c.; but though they are all distinct, they
have characteristics which depreciate them; and unless
in large establishments where they are grown for the

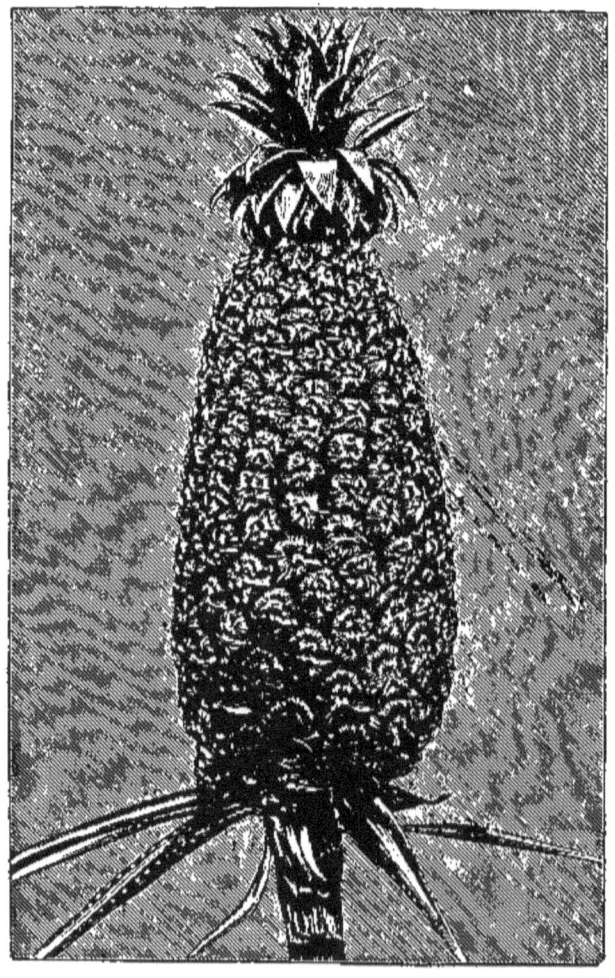

Fig. 5.

sake of mere variety, they have no claims upon the
space at the disposal of pine-growers in general; and
as I prefer to occupy space with cultural directions,

as being the more useful, I will not describe any of these varieties that I cannot recommend. In my own practice, I have found the Queen, Smooth Cayenne, Charlotte Rothschild, and Prince Albert the best and freest-fruiting.

SOIL.

Dr Lindley, in his 'Theory of Horticulture,' says,—"We are informed by Beyrick, that the pine-apple in its wild state is found near the sea-shore—the sand accumulated there in downs serving for its growth as well as for that of most of the species of the same family. The place where the best pine-apples are cultivated is of a similar nature. In the sandy plains, Praya Velha and Praya Grande, formed by the receding of the sea, and in which few other plants will thrive, are the spots where the pine-apple grows best." Although the soil in which the pine-apple is found growing in its native or wild state cannot be taken as an absolute guide, still the fact that sand is its native choice would of itself serve to teach the cultivator that a heavy clayey soil, having a strong attraction for water, is not likely to be the most suitable for the healthy growth of pine-apples. I believe that practice has set its seal to this; at least my experience leads me to recommend a fibry calcareous loam in preference to that which all gardeners know as a heavy and tenacious loam. That in which I have grown the best pines was taken from the surface of a rocky crag, and was very full of fibre. It should be collected and stacked for twelve months before it is used; and a few months before being required for potting, put into a dry airy shed, breaking it up or teasing it with the hands—not separating a particle of the fibre, but rather sifting or shaking a portion of the mouldy particles from it. It thus forms a soil with much more fibre in it than is generally used for pines, and one which the soft, rather fleshy roots of the pine seem wonderfully to enjoy. This soil is used without any addition of manure consisting of animal excrement. I consider it very undesirable to use anything that has a tendency to produce a pasty, retentive tendency in the loam, or that would rapidly hasten the decomposition of the fibrous part of it. Animal excrement has a tendency to do both, and on that account I never use it for the pine: all that is added to or mixed with the loam is an 8-inch potful of half-inch bones, and the same quantity of soot, to each barrowful of the loam. These mixtures are highly manurial, have a beneficial mechanical effect on the soil, and offer no inducement to the inroads of worms, but the contrary.

I have always observed that the most vigorous of the roots are found in the most fibry part of the ball. Besides, turfy loam, free from all slimy matter, is regarded as the best medium for supplying nourishment in a liquid state, as will be found recommended further on in this treatise. I would therefore recommend a friable loam, with all the verdure that grows on it—such as the top three, or at most four, inches of an old pasture, where such can be had; and should such not be attainable, and the cultivator therefore be obliged to use a heavier soil, I would recommend that a portion of sand, pounded oyster-shells, charcoal, old plaster, or mortar-rubbish be mixed with it, to prevent its ever becoming compressed or puttied —a condition which is most injurious.

THE PINE-APPLE.

PROPAGATION.

Generally there is little trouble in propagating and keeping up a stock of young plants, as the majority of varieties propagate themselves freely by suckers and crowns. The latter I never use, except in the case of some varieties which are very shy in producing suckers—such, for instance, as the Smooth-leaved Cayenne, and C. Rothschild. Suckers are much more desirable, and grow into strong plants more rapidly than crowns. Those varieties that do not produce suckers in sufficient abundance I always find easily enough increased by preserving the old plants from which the fruit is cut, stripping all the leaves off them, and placing them entire in shallow boxes, covering them to the depth of an inch with light rich soil, in a bottom-heat of $90°$. In this way every latent bud on the stems bursts into growth; and as soon as they begin to emit roots, they are twisted carefully from the old stem, and potted in 6-inch pots. The stems may also be split up through the middle, cut into pieces according to the number of buds, potted singly in small pots, and plunged in bottom-heat. This plan gives more labour and requires more room, and sometimes the pieces rot before the buds start. However, either way can be practised with success.

By this mode of propagation a clean stock can be produced from plants infested with scale. In this case the stems should be well scrubbed with soap and water before being placed in boxes or pots. In this way a perfectly clean set of plants have frequently been produced from stock which had been overrun with insects.

SUCKERS.

Suppose a quantity of suckers to come under treatment from the beginning of August to the middle of September—the time when suckers are generally in a fit state to be taken from plants that have produced the summer supply of fruit: let them be carefully detached from the parent plants, cut their rugged base smoothly off with the knife, and remove with the hand the short scaly leaves which cluster round their base, and under which appear the young roots. The leaves should not be removed any higher up than where these young roots assume a brownish hue. As this operation is proceeded with, the suckers, for convenience, should be classed into two lots, the smaller and the larger being placed by themselves. The larger set, presuming that they are strong and healthy, are to be potted in 8-inch, and the smaller in 6-inch pots. The pots, if not new, should be well washed both outside and inside. The crocking should be efficiently performed, using rather finely broken crocks with all dust sifted out of them. They should be arranged in the bottom of the pots to the depth of one and a half inch in the 6-inch, and two inches in the 8-inch pots. Over the crocks should be placed a thin layer of dry moss or the most fibry part of the loam, and over all a sprinkling of fresh soot, which acts as a barrier to worms and affords a stimulant to the plants.

In potting the suckers, place them sufficiently deep in the pots to keep them steadily in their places; press the soil firmly about them with a blunt-pointed piece of wood, and leave it about three-quarters of an inch from the rim of the pots, that there may be no difficulty in watering them when necessary. It being presumed that a pit was previously made ready for their reception, they should be plunged at once to the rim of the pot; and should the bottom-heat be derived from leaves or tan, or both, and not likely to exceed $90°$, the plunging material may be placed firmly round the pots; but if the heat is likely to exceed $90°$, let the material be placed lightly and openly round them. Let the plants be arranged as previously directed according to the structure of the pinery, and in doing so avoid crowding them together, the consequence of which is to draw the young plants up weakly—and to make good plants of them afterwards is almost impracticable.

They must now be shaded from the sun during the brightest part of the day for ten or fourteen days, or, in fact, till it be found that they are making roots. In the afternoon, when the shading is removed, they should have a gentle dewing overhead through a very fine rose. The shading and dewing must not be abruptly discontinued, but by degrees; and entirely given up whenever the young roots are two or three inches long. Then they should have a watering with water at $85°$ sufficient to moisten the whole ball. After this they soon begin to grow freely, and air should be given early in the day when fine. A good supply of air, as much light as possible, and a moderately moist atmosphere, with a very sparing use of the syringe only in hot weather, will prevent them from making a weakly drawn growth.

From the time the suckers are potted, the great object is to obtain a compact sturdy growth as one of the principal points of future success, which will enable the plants to go through the rigours of winter with impunity. This is dependent chiefly upon free exposure to light, a good supply of air without draught, and a moderate amount of heat and moisture both at the roots and in the air.

The night temperature for September should range from 65° to 70°, with 10° to 15° more for a while when shut up in the afternoon with sun-heat. After the middle of October the heat should be 5° less, and it should gradually decrease till, by the middle of November, it is 55° to 60° at night according to the weather, with 5° more by day. During October the bottom-heat should not range higher than 85°; and for the three following months I consider 75° quite sufficient to keep the roots healthy through these dull months. In olden times, when every sucker potted in autumn was deprived of its black and lifeless roots in spring, it was considered that pines lost all their previous year's roots in the common course of nature. But there is no doubt whatever that the real cause of the evil arose from the common rule of renewing the beds in which the pines were plunged at the fall of the leaf, the consequence of which was a degree of bottom-heat which pine-roots cannot bear and live. The good pine-grower of the present time is not satisfied if, when September-potted suckers are shifted in early spring, their roots are not white and full of life, instead of black and shrivelled.

Under ordinary circumstances I would recommend that the suckers now being treated of should be kept quiet from the middle of November till the middle of February, and not encouraged to grow. To rest them thus, a temperature of 55° is preferable to 60°, unless during very mild weather, but 60° should never be exceeded. The atmosphere should be dry rather than otherwise; and I have very rarely found that, when grown on a bed of leaves and tan, during these months they ever require any water at the root. The tan in which the pots are plunged is generally moist enough for the maintenance of pine-roots in a healthy condition, and the soil in the pots is regulated as to moisture at this season by the state of the plunging material. Where the bottom-heat is supplied with hot-water pipes in air chambers or tanks, the plants may require an occasional watering; but with the bottom-heat that I have named, the waterings required will be very few indeed. Young stock is in very little danger of fruiting prematurely from being kept rather dry, if all else be right; and in all other respects it is much the best practice.

When the thermometer rises to 65° a little air should be put on, always at the highest point of the pit or house. But, unless during a continuance of dull damp weather, the temperature should not be purposely raised in order to admit of giving air. In most pineries there is a sufficient amount of circulation going on in the atmosphere through the laps of the glass and other chinks to render systematic airgiving, with the low temperature and dry atmosphere that I have recommended, unnecessary. It is therefore only during sunny days, when the heat is raised, that air-giving must be carefully attended to during the season of rest.

Under ordinary circumstances this is the winter treatment to be recommended as that which will give succession plants in the most robust and healthy condition in spring, and that can be grown into the very best fruiting stock by the following autumn. Scarcity of intermediate plants may, however, in certain cases, render it desirable to considerably increase the size of the plants in order to gain time. When such is the case they should be kept gently on the move all winter, by keeping the temperature at from 60° to 65°, with a little more moisture at the root than has been recommended. The highest temperature named should be given during the brightest and calmest weather, when it can be secured without anything like violent firing; and during weather the reverse of this, the lowest is much the safest. This winter growth can only be pursued with success when the pineries are light and fully exposed to every ray of sunshine that can possibly be had. Otherwise the plants will become drawn and weakly, a condition which will more surely than any other defeat the object in view. It is only when there is a scarcity of good succession plants that I would advise these autumn suckers to be pushed on, with the view of resting them in April and May, in order to start them for supplying fruit in autumn.

SUCCESSION PLANTS—SPRING TREATMENT.

This is the distinguishing term which is applied in spring to the suckers of the previous autumn, and it is as succession plants that I will now treat of their spring and summer culture.

Except in the case of plants which may have been kept in a growing condition all winter, it rarely occurs that September-potted suckers require a shift into larger pots before the middle of February; more especially if at first they are potted into 6-inch and 8-inch pots as recommended. In my own practice I am, however, never regulated by dates, but by the condition of the plants. Succession pine plants in a proper condition for shifting I would describe as those which have moderately filled their pots with roots in a white and healthy state of preservation. They should not be shifted till roots have formed themselves round the ball of soil sufficient to keep it together. On the other hand, they should not be allowed to stand unshifted till they become anything like pot-bound. If the former condition is not arrived at before the middle or end of February, the operation of shifting should be deferred, and the plants gently excited into action by increasing the night temperature to 60° when cold, and 65° when mild, with 10° more with sun-heat by day. Keep the bottom-heat at 85°, and increase the moisture both in the soil and air, till their roots are in the condition I have named. Should they have become pot-bound, which sometimes occurs in the case of strong suckers, especially when in the smaller-sized pots, the balls should be partially broken up with the hand, and the roots disentangled as much as possible. Plants with hard matted balls seldom start freely into growth, and are liable to start prematurely into fruit. The best way is to keep a watchful eye on young stock and shift them the first opportunity after they are sufficiently rooted.

About a week before the shifting is performed, the plants should be carefully examined, and all those that are dry should be watered, so that at shifting time the soil may be moderately moist. If shifted with their balls dry it is difficult to properly moisten them afterwards, particularly as it is not desirable to water them immediately after being shifted. The other preliminaries of getting the necessary amount of soil prepared and placed in some place to warm it, and the pots cleansed, crocked, and arranged in convenient readiness, should be all seen to before the day on which the pines are to be repotted. Hurry and confusion will thus be prevented in taking advantage of the first mild day for shifting and rearranging the succession stock. In draining the pots it must be borne in mind that the plants are to remain in them till they have perfected their fruit and a crop of suckers for another season's stock, and the drainage should be efficiently performed, as directed when treating of suckers, only the depth of crocks should be a little greater in the case of the pots recommended for fruiting in.

The house or pit intended for the reception of the plants after they are shifted should be thoroughly cleansed. The glass and wood-work should be all washed, and the walls whitewashed with hot-lime, so that there may be admitted and diffused as much light as possible, which for a stocky and fruitful growth early in the season is one of the most important conditions in the cultivation of the pine-apple. In the case of those who are dependent on fermenting material for bottom-heat, all that may be necessary in relation to that will be to add about six or eight inches of fresh tan, well mixing it with a foot of the surface of the old bed. But should the leaves have been several years in the pit, and the heat much declined, it will then be necessary either to take out the tan and mix in some fresh leaves with the old, or to add a greater proportion of fresh tan without interfering with the leaves at all. In the latter case the old tan should be sifted, preserving the roughest part of it. There is not an operation connected with the growth of the pine-apple that I dread more than entirely renewing the leaves and tan in pine-pits; and rather than run the risk of sudden and violent fits of bottom-heat, I have allowed the leaves in the bottom of pits to remain undisturbed for six or seven years at a time. I have always found, where tan is easily got, that the safest and best way is to sift the tan once a-year, and mix in with the old a few inches of fresh tan, which raises a steady and sufficient amount of bottom-heat. A bed so managed is far more under control than when the leaves and tan are annually or even biennially renewed entirely. All this labour in preparing beds is dispensed with where the bottom-heat is supplied by a well-regulated system of hot water—and the labour connected with the shifting and arranging of pines in spring or any other season is much lessened and simplified.

Supposing that I am now treating of Queens that are required to fruit early in the following year, to supply ripe fruit in May and June—little more than eighteen months from the time they were taken as suckers from their parent plants—I prefer shifting them into their fruiting-pots at once, instead of giving them two small shifts. Indeed, the size of pots into which they have been potted as suckers, and those into which I shift them for fruiting, admit only of one shift without reducing the balls. The strongest plants in 8-inch are shifted into pots 12 inches wide and as many deep, and those in 6-inch into 11-inch pots. These sizes are sufficient for the production of the very finest pines. Fine fruit is not dependent on size of pot so much as on other points of culture. I have had fine crops in 9-inch pots, but they require more attention in watering. And what is of no small consequence, especially to those who have a regular supply of fruit to keep up from limited accommodation, it is found that pine plants grown in comparatively small pots are much more manageable in the way of getting them to start than when grown in larger pots. From this it will be observed that all that I recommend in the way of repotting pines, in their progress from the sucker state to their yielding and ripening their fruit, is simply one shift.

Before turning the plants out of their pots, a few of the short sucker leaves round their collars should be stripped off. When turned out of their pots, all inert soil on the surface of the ball should be removed with the hand, and the crocks taken from the bottom part, taking care not to injure the roots. The ball should then have a gentle tap or two with the palm of the hand, and the outside roots be disentangled a little without breaking up the ball. This is what is recommended in the case of plants that have the soil and roots in a thoroughly satisfactory condition—having fine healthy white roots, with a moderately matted ball, and the soil in a healthy condition. When, as may occur in individual plants, the soil is either over dry or soured with wet from having stood in a drip, it is best to shake out the plants either more freely than I have directed, or entirely, according as the condition named may exist to a limited or extreme extent. The pots should be filled firmly up with soil, so that the plants when placed in them may be from two to three inches deeper in the pot than they were before. Being an advocate for very firm potting, I recommend that the soil should be rammed firmly round the ball with a blunt-pointed piece of wood. Be it remembered that the soil I have recommended to be thus acted upon is not a damp mixture of heavy soil and animal excrement, but a light turfy loam through which water passes freely; and the more firmly it is put into the pot the less water it holds in suspension, a point of no small importance in the growth of so succulent a plant as the pine. I never remember seeing really healthy pines or fine fruit in a rich puttied soil, holding a superabundance of water about the roots. The soil should be made thus firm all round the ball and about the collar of the plants up to within an inch of the rim of the pot.

When the whole are shifted they should be plunged in their growing quarters at once. And should there for the time be a scarcity of room for the desired number, with the prospect of more room in the course of a few weeks by getting rid of others that are fruiting off, they may be arranged rather thicker than is proper for them to make their summer growth. But if at once they can have the necessary amount of room —namely, two feet from plant to plant in and between the rows—all the better; for there is nothing more to be deprecated in pine-growing than overcrowding.

Particular attention must now be paid to the bottom-heat; 85° to 90° should be aimed at. And, where the heat is derived from tan and leaves, should it exceed 90°, the pots should be moved from side to side, so as to leave an opening round their sides. Although there may not be absolute danger of burning the roots while they have not reached the sides of the pots, yet too much bottom-heat causes an over-rapid growth at too early a season, which, in the absence of longer days and brighter sunshine, is exceedingly undesirable. During the month of March the atmospheric heat should range during cold dull weather from 60° to 65° at night. I am not particular as to a few degrees, but much prefer being guided by the outside temperature. During bright sunshiny days, when the pinery can be shut up in the afternoons with sun-heat, the temperature at 8 P.M. may be 70°, allowing it to sink to 65° by morning.

For a few days after being shifted keep them rather close, and the atmosphere moist, till they begin to lay hold of the fresh soil. Then give a little air daily as soon as the temperature exceeds 70°; and with steady sunshine the amount of air may be gradually increased till 2 P.M., when it should be gradually diminished according to the character of the day, and shut up so as to run the heat up to 80° for a short time before dark. There should not be any attempt at causing a rapid growth till the days get longer and the light more intense. The plants will root freely into the fresh soil, from the increased bottom-heat and the healthy irritable state of the roots, without much perceptible top-growth for a time.

There will not be any necessity for water at the root for some time—not, certainly, till the early part or middle of April, and even then water should not be over liberally supplied. The experienced can tell by the very appearance of the plants when they require it; but the inexperienced should examine the soil occasionally and apply water when it becomes dry a few inches from the surface of the ball. Rain-water is of course the best, and it should be heated to not less than 80°, nor more than 85°. At this season it is much safer to err on the side of giving a moderate amount of water than to keep the soil too wet while it is yet unoccupied with roots. The perspiratory organs of the pine are not very active at any season; and as the plant partakes so much of a succulent nature, a little extra moisture in the air is a much safer way of preventing injury from drought than by applying much water at the roots so early in the season.

THE PINE-APPLE.

It is often found, in the case of those who have next to no experience in pine-culture, that young pines after they are shifted are kept far too wet. I have taken the soil out of the pots and squeezed the water out of it. No more fatal course can be pursued at any stage of their growth, but particularly in spring when newly shifted.

SUCCESSION PLANTS—SUMMER AND AUTUMN TREATMENT.

Raise the night temperature by the end of April to 70° when the weather is dull, but when the pineries can be shut up with sun-heat the thermometer may range to 75° at 10 P.M. with advantage, falling to 70° towards morning. With a proportionate amount of atmospheric moisture the plants will now begin to grow freely. The increase of light and sun-heat will render a less amount of fire-heat sufficient, and, as a general rule, the state of the weather admits of a more liberal supply of air being given. This enables the cultivator to push forward his early plants without the danger of drawing them, which exists at an earlier period of the year.

In order to keep up the temperature with as little fire-heat as possible, air should be given early in the morning, almost as soon as the sun strikes the glass, and increased as formerly directed, so that the shutting up may take place at an earlier hour than is usual. This allows of the maximum temperature while there is yet a strong light, and husbands the heat of the sun for the evening. The steaming-troughs should be filled up every day when the pinery is shut up, and at the same time the paths and walls damped with the syringe. Without a moist atmosphere at this season the growth will be deficient in broadness, texture, and that dark-green hue which indicates that all is going on well. I disapprove of heavily syringing young growing pines, and much prefer the moisture to be applied by evaporation. On the afternoons of very bright days an occasional syringing overhead through a fine rose is beneficial, and keeps the plants clean; but regular heavy syringings have a tendency to keep the soil in a puddled state, as the leaves conduct all the water that falls on them into the pot, and this has a tendency to produce a soft unfruitful growth.

With increased air, light, and heat, and the very moderate syringings recommended, the state of the soil as to moisture must be carefully watched. An equal and healthy amount of moisture must be maintained. No amount of attention should be considered too much to prevent the soil from becoming dusty-dry on the one hand, or over-wet on the other, otherwise a check may be given and an amount of mischief produced that no after-treatment can retrieve. It is a great mistake to suppose that a check is not as likely to arise from plants being kept too dry as from the opposite extreme.

When bottom-heat depends on leaves and tan, it not unfrequently occurs, although the heat may be just right in March and April, that the hotter sun of May causes an increase of heat just at a time when the young roots are reaching the sides of the pot and are most susceptible of injury. The safest way is to have a thermometer in the bed, and as soon as the heat exceeds 90°, to shake the pots from side to side and leave an opening all round them for the heat to

THE PINE-APPLE.

escape. After the heat subsides, the tan can be pressed to the sides of the pots again. Of course, when bottom-heat is derived from hot water, it can be easily regulated without these precautionary measures, which apply only to fermenting materials.

The temperature should now be carefully regulated, and fire-heat applied in the evening just in time to prevent the heat from sinking below 70° at 10 P.M. And when the morning gives signs of a bright day, the fires should be damped down the first thing, and kept low all day. There is nothing more injurious than to have hot pipes, and a bright sun, with a maximum supply of air on. Such a state of things creates currents of scorching dry air, very trying to the plants, and robs the pineries too much of moisture. By the middle of May the plants will be growing freely, and moisture and air must be increased in proportion to the progress they make. The house should be damped the first thing in the morning as well as at shutting-up time. And after being shut up close for four or five hours, when the weather is calm and very warm, a little "chink" of air should be left on all night. A little more air should be put on at 7 A.M., and gradually increased with the rising of the sun, till at twelve o'clock there is sufficient to create a circulation among the plants. Air should be given at the back or highest part of the house or pit; but, unless when the weather is close and sultry, none should be given at the front. With the increase of heat, light, and air, they will make rapid progress, and consequently more water at the root will be required, and it should always be about the same temperature as the bottom-heat. I have found Peruvian guano the best and most convenient stimulant for mixing with the water—not in strong doses now and then, but simply to well colour the water with it every time the pines are watered: an ordinary handful to four gallons of water is sufficient.

In some localities, and with fine summer weather, after midsummer the temperature can often be kept up sufficiently without the aid of fire-heat. In a close structure there will be no difficulty in doing so, especially when early air-giving and shutting up is practised. The heat can thus be husbanded so as to keep the thermometer at 75°; and when this can be accomplished without the aid of fire-heat, so much the better in all respects. This is, I am aware, not applicable either to all localities or all seasons; for many climates, even in favourable summers, will render the use of the fires necessary the whole season.

Although very much opposed to shading pines in a general way, it is sometimes necessary, when they are growing rapidly and the weather becomes suddenly very bright after a continuance of dull weather. The shading should never be heavy nor long continued. Tiffany or hexagon netting I have always found sufficient, and that only during the brightest part of the day. If all is going on right at the roots, and a moist atmosphere is steadily kept up, I have never found a necessity for more shading than this. At the same time, it is most undesirable that pines should become browned and wiry; and slight shade and more frequent gentle dewing at shutting-up time should be resorted to as soon as signs of this appear. Of two evils, the browning of the leaves is not so injurious as a weak watery growth—the result of too much shade and a close atmosphere. I find the Smooth-leaved Cayenne much more impatient of sudden bursts of

THE PINE-APPLE.

bright sun than Queens or other varieties; and to grow it to perfection it should never be allowed to become much browned. In the case of this fine variety I have in bright warm seasons fixed a single ply of hexagon netting over the pits, and allowed it to remain for a couple of the hottest months. This simply breaks the power of the sun a little. In order to prevent this wiry, browned condition during summer, care should be taken that the plants are never once allowed to go too long without being watered, and a uniformly moderate moist state of the soil must be maintained.

Should any of the plants throw up young suckers from the axils of the lower leaves, they should be removed at once. The best way of doing this is to have a long-handled pair of broad-mouthed pincers, with which the suckers can be easily twisted out as soon as they are observed. Where much syringing overhead is practised, suckers frequently show themselves in abundance, in the case of Queens particularly. This is one of the many evils which result from the too liberal use of the syringe. It often occurs during the season of rapid growth that some of the centre leaves adhere closely to each other for a longer time than is good for them: they should be separated either with the hand, or with a slight touch of a stick where the hand cannot reach them.

As the stock of which I am now treating consists principally of plants that are selected to start into fruit for the early supply of next season, the plants should always have their pots well filled with roots, and be of a stocky well-matured growth, by the end of August, otherwise there is little certainty of their being got to start in time to be ripe in May and June.

If grown on the shady, large-pot, and wet-at-the-root system, they will not be in a fit state for the purpose now named; and even with the best of management to induce them to start without first making a growth in January and February, it is necessary that they should complete their growth early under the influence of plenty of light and air, or they will make a fresh growth when the temperature is raised with the object of starting them, instead of coming up at once into fruit. True, those which make a growth first, I have always found, throw the finest fruit; but where an early summer supply of fruit is required, it must be had from those which start without any growth. In properly preparing plants for this purpose, there are two things which must be guarded against. The one is that of having the plants pot-bound too early, and subjected to a high temperature too long in autumn. In this case the fruit comes up slowly late in autumn, or in winter, a hardened knot like a thimble, and is worthless, especially in the case of Queens. The other is a watery immature growth, from which it is impossible to get early fruit.

In September water must be judiciously and very sparingly applied. No more should be given than is just sufficient to prevent the plants from suffering either from aridity of atmosphere or dryness of soil. Give a liberal supply of air on fine days. Towards the end of September they should be as completely at rest as a comparatively low temperature, a dry atmosphere, and a proportionately dry state of the soil in which they grow, can place them. I have frequently allowed Queens in this stage to remain without a drop of water at the root from the first week in October till January, and found the plants so treated in the very best condition. To start pines into fruit at any given time, and more especially very early in the year, it is necessary to their doing so satisfactorily, that they have a period of rest previous to their being subjected to the treatment required to start them. Such as have completed their growth as I have described early in the season, can have from ten to twelve weeks' rest, and be started in time to ripen their fruit in the end of May and June. From the beginning or middle of October, onwards to the end of December, it rarely occurs that pines intended to start thus early are the better for a drop of water, when grown on a bed of fermenting material. And when the bottom-heat is supplied with pipes, it is much the safer way to keep the plunging material moderately moist than to water the pines often.

The night temperature should drop gradually to 60° by the middle of October. In November, and until the time they are to be started, I prefer the temperature at 55° at night during cold windy weather, and 60° when mild. The bottom-heat should be proportionately low, just enough to maintain the roots in a white healthy condition, and 80° is quite enough for that. When with sun-heat during the day, which may occur during clear frosty weather, the temperature exceeds 65°, air should be given. With such weather as this it is sometimes necessary to fire sharply at night to keep up the required temperature; in which case the fires should be checked the first thing in the morning, especially when a cold night is succeeded by a bright day. Where it can be so arranged that covering can be used over the glass during cold weather, it prevents radiation, and the atmosphere can be kept in a condition much more congenial to pines than when more fire-heat is necessary. For although a damp atmosphere, which leads to an accumulation of moisture and to drip, is by all means to be avoided at this season, yet a parchingly dry atmosphere produced by highly-heated pipes is very prejudicial, and cannot well be counteracted in winter without producing the opposite evil. Hence the benefit of covering the glass at night. When, however, it becomes necessary to apply moisture to counteract the too drying effects of hard firing, the best way is to sprinkle the paths instead of the pipes, because the moisture will be carried more gradually into the atmosphere, and is therefore not so likely to accumulate and drop into the centres of the plants, which, as all pine-growers have doubtless found out, is attended with spotted leaves, and not unfrequently deformed fruit.

Winter treatment the reverse of what I have here recommended—a high temperature and more water at the root and in the air—causes the plants to grow all winter; and from want of light and air they become drawn and weakly—in fact, worthless,—or probably some of them may start at the dead of winter, when, particularly in the case of Queens, there is very little chance of their blooming and setting properly, and will either way be worthless. An instance of such treatment once came under my notice, when, instead of a low temperature, 75° of heat was kept up during the whole resting season, with moisture in abundance. The consequence was, that when the time for starting them came round they were tall, tender, and only fit for the waste-heap.

Pine plants arrived at the stage I have been now

treating of are termed fruiting plants, and under that heading I will speak of their further treatment.

FRUITING PLANTS.

Ripe pines being required in the early part of June, it will be necessary to set a quantity of Queens in motion by the first of January, to succeed those which are generally termed winter and spring fruiters, and which will be treated of by-and-by. Queens are by far the best variety to start at this season, with the view of getting ripe fruit from them quickly to keep up the succession after the winter fruiting varieties. Yet for the sake of variety, and also to keep up as long a succession as possible from the same lot of plants, it is desirable to start a few of the later varieties at the same time; but Queens should form the great majority.

Where bottom-heat is derived from leaves and tan, the bed in the fruiting pinery should have fresh material added to it, as formerly directed, to increase the heat to from $85°$ to $90°$; but in doing this, very particular attention must be paid to the state of the bed, as over-much bottom-heat at this stage would prove fatal to anything like success. The principal part of the roots being at the bottom and round the sides of the pots, they are now more than ever particularly liable to suffer from too much heat, and great caution is necessary. Should there be any fear about the over-heating of the bed after it is prepared, it will be much safer to only half plunge the pots at first, till it be certain that the heat will not exceed $90°$.

Those who have the more desirable and superior

appliance of hot-water pipes or tanks for bottom-heat, will be spared the trouble and anxiety which is attached to the otherwise by no means inefficient, when well managed, fermenting bed. They can regulate the bottom-heat with much more ease and safety.

In selecting the plants for starting at this early season, those only should be taken which are most likely to start without making a growth. I will therefore suppose that the cultivator has a hundred plants of those treated of as "succession plants," and that from these it is desired to have a supply of ripe fruit from the first of June till October, and recommend that fifty of those most likely to start at once should be selected. In doing so the experienced eye will fix upon those with the thickest collars, and that have the greatest number of short sharp-pointed leaves, thickly set together in their centres. These are the most likely to send up their fruit without making a fresh growth, although some of them may disappoint even the most experienced; still, in a general way, when prepared the previous autumn and winter as I have described, they will not disappoint.

In arranging and plunging these plants, a few of the bottom leaves should be stripped off, all the loose soil on the surface removed, and a top-dressing of loam put on, pressing it firmly to the collars of the plant and the sides of the pot. In moving these plants it is a common practice to tie the leaves up for the sake of convenience; but I would here say that it is a practice that is injurious at any stage of the pine's growth, and particularly when the plants are full grown, and should have stubby, short, thick leaves that will not bear being squeezed into a bundle without considerable injury. I seldom tie pines up at any stage when working amongst them. Those who shift and plunge the strong prickly varieties can easily protect their hands from being torn by wearing a pair of gloves. In plunging them they should not be put thicker than two feet from centre to centre, and that side of the plant which has been to the sun all the growing season should be placed so still. Indeed, very strong plants require more room.

As soon as they are all plunged, if they are dry, water them with guano-water at 80°, giving them sufficient to moisten the whole ball, but be careful not to splash it about the leaves. The atmospheric temperature for January should be 65° at night, and 70° by day without sun; with sun, 80° will be sufficient, and air should be given when it exceeds that. The moisture in the air must also be proportionately increased, and should be done by sprinkling the paths and walls with tepid water two or three times a-day, instead of steaming the pipes for the present. A watchful eye must be kept on the state of the soil, and no more water given than is sufficient to keep it moist, but not wet. With too much water, and the degree of top and bottom heat now necessary, the tendency of pines to make growth at this season and miss starting for the time being is increased. With these conditions the plants having a mass of healthy roots in an irritable state will soon show signs of motion, and all the more surely in proportion as the heat and moisture are steadily administered.

In February the heat must be advanced to 70° at night, and 75° by day, and air put on when it exceeds 80° with sun, shutting up the house early in the afternoon so as to husband sun-heat. The moisture in the air must not be much more than in January, and the same cautious application of water to the roots must be observed till the fruit makes its appearance. Most of the plants will show fruit before the last week of February. The centres of the plants will be observed to open by degrees, and on examining them the young fruit will be found emerging from the centre. Whenever this is observed, the plants, if inclining to the dry side, should have a watering sufficient to thoroughly moisten the whole ball, and the bottom-heat already named should be steadily kept up.

Supposing all the plants to have shown fruit, the night temperature for March should not range under 70° nor over 75° with the mildest weather. There being generally great fluctuations of weather during this month, the temperatures I have named should be aimed at accordingly. The moisture in the air must be sparingly applied till the fruit is out of flower, and air admitted on all fine days, putting it on early in the morning, and shutting it off early in the afternoon. Water at the root will be more frequently required, especially when they are plunged over a hot-air chamber. But avoid, as one of the greatest possible evils, a wet sloppy state of the soil. As soon as they are out of flower, sprinkle them overhead every fine afternoon with clear water at a temperature of 80°. As the season advances, with longer days and shorter nights, early shutting up with sun-heat must be practised; but, except with sun-heat, I do not recommend in April any increase of night temperature over that recommended for March, even though it be required to ripen the fruit with as much speed as possible. The forcing should be accelerated by day with sun-heat. Shut up soon after three o'clock, giving them a gentle dewing overhead, filling up the steaming-trays, sprinkling the surface of the plunging material and about the collars or bottom leaves of the plants. The temperature may then be run up to from 85° to 90° for an hour or two. The fires, which should now be low through the day, should be quickened in time to keep the heat from falling below the proper night temperature at 10 P.M.

Under this treatment the fruit will swell rapidly, and careful attention must be paid to watering. The great thing to be aimed at being to keep the soil in a healthy growth-giving state—moist, but not wet—it is a common practice to give occasional strong waterings with guano, sheep, or deers' dung. Instead of this, I prefer, as already directed for succession plants, to water every time with a weaker solution of these manures, and I prefer guano to any other; and during the rapid growing season, I always put a little of it into the evaporating pans once or twice a-week, and find it gives that fine dark-green hue and thickness of texture so desirable to see in pines. They should be gone over as soon as the suckers appear, and where there are more than two to a plant remove them. When suckers or gills appear on the stems or under the base of the fruit, they should be removed immediately they are discovered.

The month of May generally brings comparatively warm sunny weather, and vegetation gets into full play; and I am not sure but what May is the very best month in the whole year for swelling off pines. It is not generally so hot and scorching as the succeeding three months; less air is therefore needed. The pineries can be shut up earlier, so that less evaporation goes on, and the swelling fruit can have a longer period of sun-heat and moisture in the afternoon than when the sun is more powerful, and when it is not safe to damp and shut up before four o'clock. Advantage should therefore be taken of these circumstances, and the fruit pushed on, when it is an object to get them ripe as soon as possible. Under these circumstances, the heat may be run up to from 90° to 100° for an hour or two, and the air loaded with moisture. Syringing must not, however, be to excess, or the result will be large crowns and an undue growth of suckers, to the detriment of the size and appearance of the fruit.

When the fruit begins to change colour, which, if the plants have been set agoing in January, will be in the end of May or early in June, it is necessary, in order to get highly-flavoured fruit, to increase the amount of air, and decrease the moisture both in the air and the soil. Indeed, as soon as the fruit is half coloured, no more water should be given than is necessary to keep the plants from suffering, and the moisture of the atmosphere should be gradually withdrawn. At the same time, avoid starving them into maturity.

RETARDING AND KEEPING PINE-APPLES AFTER THEY ARE RIPE.

When a greater number of pines begin to ripen at any given time than is necessary to supply the demand, it then becomes desirable that a portion of them should be retarded to form a succession of fruit in good condition. In the absence of a compartment specially for the purpose, I have frequently placed them in a vinery where grapes were nearly ripe, and

THE PINE-APPLE.

where the temperature was comparatively cool, with a circulation of dry air. In such a place, pines that have begun to colour ripen slowly, and they are excellent in flavour. The cool dry air of the vinery, and the shade of the vines, are good retarding conditions; and this is as good a way, apart from having a place for the purpose, as any that I have tried. I have also removed them to a cool dry room when about half coloured, and kept them there a month or six weeks, and found them in excellent condition. This treatment, of course, applies to summer fruit. Later in the season I have kept Smooth-leaved Cayennes in a room for six weeks after they were quite ripe. In this way a succession of fruit can be very much extended as compared to keeping them in a warm pinery.

When the fruit is all cut from a pit or houseful of plants, the suckers should be carefully attended to. The comparatively dry condition of the air and soil which is necessary to good flavour is not favourable to the suckers at this hot season of the year; consequently, when the suckers are strong, I frequently detach them from the plant as soon as the fruit begins to colour. If the suckers are small when the fruit is cut, they should be left on the parent plant; then the soil should have a good watering to encourage them to make further growth. It rarely occurs that they are not quite large enough to be potted about the time the fruit begins to ripen. I may here remark, that the practice of allowing the suckers to lie in a cool dry place, with the object of what is called drying them, is one for which I never could see any reason, or any good end that could be gained by it. On the contrary, in my opinion, the practice is injurious to the progress of the young plants. To say the least of it, it is attended with a loss of time.

. When it is desirable to have the fruiting plants of which I am now treating to ripen earlier than the beginning of June, they must, of course, have heat applied to them in December instead of January; and with properly constructed and heated pineries there is nothing to prevent this. But where the houses are not light, tight, and well heated, it is a matter of no small difficulty, and it is much safer to wait for the "turn of the day." The other half of the set of fruiting plants of which I have been treating should be kept quiet till the end of February. Introduced into heat, and managed in the same way as the early half, they will come in as a succession lot. And, as is always likely, a good many of them which the experienced eye rejected while selecting the earliest, make a growth before starting, and in that way still further lengthen out the succession of ripe fruit from this portion of the stock. For this purpose Queens are most useful in all respects, and can be had in good order from May till November.

I have considered it the best way to follow out the treatment of this one set of plants, without mixing up with their management that of different sets of plants necessary to supply ripe fruit in winter and spring. Of these latter I will now speak.

HOW TO KEEP UP A CONSTANT SUCCESSION OF RIPE FRUIT ALL THE YEAR.

Where a regular supply of fruit has to be kept up with the least possible intermission all the year round, it is more certainly accomplished by potting a quantity of suckers at frequent intervals. Supposing that a number of suckers are potted August 1880, these will give the earliest fruit for 1881. And those that ripen in September and October, give the suckers that will succeed the earliest lot, so that these two sets supply fruit for six months of the twelve. The other six months of winter and spring—particularly spring—are those in which pines are most valued, as other fruits are then scarce. March and April are the most difficult months of the whole year in which to have ripe pines.

In June and July I always endeavour to start a quantity of the Smooth-leaved Cayenne and Charlotte Rothschild. These are noble pines when well grown, being unsurpassed for appearance and long keeping after they are ripe, and swell better after October than any other pines I know. Smooth Cayenne I consider the better of the two. The Black Jamaica is also a most useful pine for winter swelling, and probably is unsurpassed for flavour at the dullest season of the year. The Queen is comparatively worthless as a winter pine compared to these two; it does not swell kindly, and is always dry and juiceless compared to them.

There should be two sets of these winter sorts, as recommended in the case of Queens and other early sorts for summer and autumn fruit. The Smooth-leaved Cayenne is so very shy in making suckers that I always endeavour to save as many crowns as I can, and take all the suckers that can be got in October from the fruiting plants, whether the fruit be ripe or not. These suckers and crowns are potted generally into 6-inch pots, and shifted in spring as soon as sufficiently rooted, as described in the former part of this treatise. They are shifted into 11-inch pots, and grown on in the usual way, only that they are not kept so dry in autumn and winter as is desirable for early starting plants. The temperature, too, is kept five degrees higher than for Queens at rest; the object being not to mature the growth of these so as to predispose them to start in spring. The heat is quickened, both top and bottom, in February, and they make a spring growth; are rested in May and June by being kept drier and cooler; and then, with increased heat and moisture, I rarely ever fail in starting them all in June and July. Care must be taken that they never get too dry at the root, particularly in spring, as that would be likely to start them before they are required. This applies with the same force to Jamaicas and Charlotte Rothschilds. These will keep up the supply of fruit till the end of the year.

It is necessary to have a later lot of these varieties to come in for spring, and this I find rather difficult in the case of the Smooth Cayenne. It makes suckers still more tardily from late plants. The method I generally adopt is to save the old stems of those that ripen their fruit through the winter, and place them in strong bottom-heat to spring the latent buds. These grow into nice plants, ready to shift into 8-inch pots in September, and I shift these into their fruiting-pots in March, and by pushing them on they start in September and October, and succeed those started in June and July. For this purpose I most decidedly give the preference to the Cayenne; and from plants of it so managed, I have had very fine fruit in the spring months. They are kept on at a temperature of from 60° to 65° all winter, with a steady bottom-heat of 85°. I have frequently had ripe fruit from 4 to 6 lb. in 9-inch pots from last year's suckers.

There is nothing peculiar in the management of these

winter fruiting sorts, except it be that I never keep them so dry and so completely at rest in winter as those intended to start early. This is with the view of their not resting and maturing themselves so thoroughly in autumn and winter as would cause them to start when excited in spring. The Smooth Cayenne requires more moisture at the root when growing than is good for most other sorts. It is also more impatient of bright sun early in the season than any I know, more especially if kept gently on the move all winter. And rather than allow the foliage to become bronzed, shade should be applied for a time, as already directed. When swelling off in winter, water at the root will of course not require to be so frequently given as in summer, and there should be no syringing. The evaporating trays will keep the air sufficiently moist. Air must be put on for a short time in the middle of every fine day.

PLANTS THAT MISS FRUITING.

It not unfrequently occurs that a few plants miss starting into fruit along with the others, but continue to grow, in spite of every effort to make them fruit. The common practice is to throw these away. When I have room to conveniently operate on these, I cut the plants over at the surface of the soil, and strip a few of the leaves off them, and pot them deeply and very firmly in fruiting-pots. They are slightly shaded for ten days, by which time, with a brisk bottom-heat, they begin to send out wonderfully strong roots, and then the shading is discontinued, and they are watered. In this way they are transformed into *dwarf* strong plants, and I always find that they start into fruit very soon after, and swell off fine fruit. When I have found a set of pines that have been drawn and are not likely to be got to fruit satisfactorily, I have treated them in this way instead of throwing them away, as is often done in such circumstances.

THE PLANTING-OUT SYSTEM.

Although I have given a good deal of attention to the planting-out system of pine-culture, and made myself acquainted with the most successful instances of its adoption, I have very seldom adopted it. Not that I suppose fine fruit are not produced by it: facts prove the contrary. But with the space at my command I have decided that, to keep up the supply which I have produced nearly every week in the year, I could more certainly do so on the pot system than by having the plants planted out in beds. Plants in pots are entirely under control at all times, for being moved or removed to force forward or retard the ripening of fruit as circumstances demand. This is of vast importance where the space in pine-beds is small in proportion to the demand for fruit, and in this respect pines in pots give an advantage over the open bed. Neither do I consider it necessary to have finer fruit than can be produced from 9, 11, and 12 inch pots. In fact, it is not the size of pot, nor the greater range that the planting-out system gives to the roots, that are the principal points of good pine-culture.

The planting-out system may be practised either over a bed of leaves or with hot water for bottom-heat. The best example of this system that I have ever seen was at the Royal Gardens, Frogmore; and there, a bed of leaves for bottom-heat is preferred to hot-water pipes. The suckers are not potted, but planted at once into beds of soil over a bed of leaves about two or three feet deep. From the sucker pits they are transplanted into the succession pits, and from the latter into the fruiting pits, where they are planted two feet apart in the rows. In other respects the treatment is the same as for plants in pots.

Others again, where the bottom-heat is derived from hot water, do not have recourse to regular transplanting, but either move the stools as the fruit are cut, and put in a little fresh soil and another plant; or they adopt the "Hamiltonian system" of leaving a sucker, and sometimes two, merely cutting down the old plant to the sucker and putting some fresh soil round it. The system can of course be modified as circumstances will allow; but from all that I have seen of it, it is my opinion that as fine fruit are produced in pots; and for rapid and certain fruiting, and where the most is to be made of space in keeping up a supply, the pot system is the best. At all events, any one who makes himself master of pine-apple culture in pots can have no difficulty in growing them in open beds of soil. The same points must be aimed at in both systems. And for beginners, any errors or mistakes in management can be more easily retrieved, I should say, in the pot than in the planting-out system.

INSECTS TO WHICH THE PINE IS SUBJECT.

White Scale.—This is the most destructive and formidable insect which the pine-grower has to dread; and in forming a collection of pines, every possible precaution should be taken to avoid getting plants infested with white scale. A very few of it will soon overrun a whole collection, and cause a great deal of trouble and expense in getting rid of it. It is an oval-shaped insect, grey, speckled with brown, and adheres closely to the surface of the leaves, and preys upon the juices of the plants, rendering them very unsightly, and weakening them with great rapidity. It increases with amazing rapidity, and yields only to the most severe and laborious treatment. I have known collections which have soon been rendered all but useless through the introduction of a single plant with a breed of this scale in it.

I am glad to say that I have been fortunate hitherto to escape having anything to do with it, and have so far the want of experience in destroying it. Many are the remedies which have been recommended for its destruction; while some have looked upon it with despair, and have got rid of it only by getting a clean stock of plants, after having destroyed the infected ones, and thoroughly cleansed their pineries.

Brown Scale.—This insect sometimes affects pines, but it is not nearly so difficult to deal with as the white scale. I know from experience that syringing with clean water, heated to 140°, completely kills it without injuring the plants.

Mealy Bug.—This is also a most formidable insect to get rid of when it is established on pine plants. The white dusty material with which it surrounds itself completely protects it from the influence of hot water applied through the syringe, and it is second in its destructive effects and difficulty of being eradicated only to the white scale itself. If allowed to go on, it affects every part of the plant—the fruit, leaves, and roots. Consequently, the first appearance of it should be dealt with as a serious evil, to be checked and eradicated at once.

THE PINE-APPLE.

The most effectual remedy for all these insects is to mix four wine-glassfuls of paraffin-oil with four gallons of water, keep the whole well mixed, and apply it to the plants with a common garden syringe. Allow each plant to stand a few minutes, and then syringe freely with clean water. This destroys the insects without injuring the plants.

www.ingramcontent.com/pod-product-compliance
Lightning Source LLC
Chambersburg PA
CBHW021813220426
43662CB00006B/302